Understanding the Elements of the Periodic Table™

BERYLLIUM

Rick Adair

rosen publishing's
rosen central®

New York

For my wife, Susan

Published in 2007 by The Rosen Publishing Group, Inc.
29 East 21st Street, New York, NY 10010

First Edition

Library of Congress Cataloging-in-Publication Data

Adair, Rick.
Beryllium / Rick Adair.—1st ed.
     p. cm.—(Understanding the elements of the periodic table)
ISBN-13: 978-1-4042-1003-5
ISBN-10: 1-4042-1003-2 (lib. bdg.)
1. Beryllium—Popular works. 2. Periodic law—Popular works.
I. Title. II. Series.
QD181.B4A33 2007
546'.391—dc22

                                                      2006008492

Manufactured in the United States of America

On the cover: Beryllium's square on the periodic table of elements. Inset: The atomic structure of beryllium.

Contents

Introduction

Beryllium (Be) played a major role in launching the atomic age when it was used to discover a subatomic particle. At first, nobody had any idea beryllium could be useful in this way. A clue came in 1930 while scientists Walther Bothe and Herbert Becker in Berlin, Germany, bombarded a thin piece of beryllium foil with radioactive emissions from polonium (Po) to see if they could smash its tiny atoms to bits.

Crashing things into each other to see what happens is a favorite pastime of atomic scientists. In the subatomic world, where things are too small to directly view, this is a good way to figure out what an atom is made of and how it is put together. For example, by the 1930s, scientists had used the method to show that an atom has a central nucleus, that positively charged protons occupy the nucleus, and that smaller, negatively charged electrons orbit the nucleus.

In the case of their beryllium sample, Bothe and Becker were hoping to see protons come flying out of the foil. An earlier try at this experiment by other researchers didn't have that outcome. Bothe and Becker hoped the emissions of their more energetic polonium source would do the trick.

What they saw perplexed them. Rather than protons, the foil emissions appeared to have no electrical charge and could penetrate thick

Beryllium is a dark, brittle metal that is not found free in nature. High temperatures and strong chemical agents are needed to extract the metal from its ores.

slabs of lead (Pb). The scientists decided the "beryllium rays" must be the matter-penetrating electromagnetic energy called gamma rays and not particles. But they were not certain about this. They also reported that bombarding other elements produced similar emissions, but none were as strong as those from beryllium.

In the next year, 1931, Irène Curie and her husband, Frédéric Joliot, tried the experiment themselves and found the same results. They also found that the emissions could knock protons out of paraffin. Like Bothe and Becker, Curie and Joliot decided the emissions must be gamma rays. However, also like Bothe and Becker, they weren't completely sure.

James Chadwick at the Cavendish Laboratory in Cambridge, England, finally solved the mystery: the emissions were neutrons! He knew the instant he read about Curie and Joliot's results in mid-January 1932. There was a growing realization among a few physicists that neutrons probably resided in the nucleus, which was more massive than could be explained by the number of protons it contained. For years, Chadwick had been sure such particles existed, but he could think of no way to reveal them. His latest attempt had been in 1931, after hearing about Bothe and Becker's results, but that had failed. Now, reading the Curie-Joliot paper, he realized that the radioactive source used in his 1931 attempt had been too weak. So, equipped with beryllium foil and what he called a "beautiful source" of radioactivity, he started the most intensive research of his career.

After a concentrated three-week effort that ended in mid-February, he had shown, without doubt, that the beryllium rays were streams of the long-sought neutron, the missing piece of the atom puzzle. Chadwick's discovery marked the start of nuclear physics. So important was his finding that only three years later he received a Nobel Prize in Physics for it.

Chapter One
What Is Beryllium?

Beryllium is a gray, brittle metal unfamiliar to most people. It is a rare portion of Earth's crust, representing only about three or four atoms out of every one million. However, it is 30,000 times more common in rocks than gold. It also forms an important part of more than 100 minerals.

The difficulty and expense of extracting beryllium from rocks are the main reasons for its obscurity. Despite this, applications for it have been growing since the 1930s. It is prized for its light weight, durability, strength, and heat capacity. Beryllium has remarkable qualities, even when combined in very small amounts with other metals to form alloys. It is used in jets, gyroscopes, and jewelry, as well as in cameras, cell phones, golf clubs, watches, space telescopes, X-ray systems, trumpets and flutes, and nuclear reactors.

On the darker side, breathing dust that includes any form of beryllium can cause a devastating, sometimes fatal lung disease. Beryllium was also used to trigger the first atomic weapons and is now part of more advanced nuclear weapons.

The Discovery of Beryllium

Beryllium was discovered in 1797 by French chemist Nicolas-Louis Vauquelin. He had been asked by French physicist René-Just Haüy to see if emeralds and the mineral beryl were basically the same thing. Haüy had recently developed a mathematical theory of minerals based on the angles between facets on a crystal. The facet angles and ability of emerald and beryl to split a beam of light into two beams indicated that the two minerals were the same.

Haüy was not the first to see this similarity. Some 1,700 years earlier, Roman naturalist Pliny the Elder noted common features when comparing emeralds to "beryllus," the Greek name for a mineral that had long, prismatic crystals with six-sided cross sections, a mineral we now call beryl.

Emeralds were probably first found more than 2,000 years ago in Egypt and mined by Romans. A hundred years later, these were the emeralds Pliny compared to beryl.

Vauquelin's chemical analysis of a South American emerald confirmed Haüy's hunch: the two minerals were the same, except that emerald had small amounts of chromium (Cr), an element Vauquelin had discovered earlier in 1797. Just as astonishing as finally answering a question of identity that had persisted over centuries, he discovered yet another element in the minerals. Others had missed it because its chemical behavior resembled aluminum's (Al), which together with silicon (Si) and the new element were beryl's main ingredients.

The distinguished French chemist Nicolas-Louis Vauquelin (1763–1829) discovered the elements beryllium and chromium. He was also the first to recognize an amino acid, which is the main component of proteins, and did groundbreaking work on the chemistry of the brain.

He announced his finding the following year, in 1798, saying he had found a new "earth" element contained in the beryl. (Earth elements don't dissolve in water.) "Beryl earth" was translated into German as "beryllerde." *Erde* is German for "earth." Later, when chemists realized the alkali earth "elements" were actually compounds, they coined new words using the Latin *-ium* ending to indicate the element, "beryllium" in this case. However, others, particularly the French, called Vauquelin's discovery "glucine," from *gleukos*, the Greek word for "sweet," because some compounds of the new element are sweet. Just as "beryllerde" became "beryllium" when the element was isolated, "glucine" became "glucinium," or "glucinum," names that stuck until 1949, when they were rejected in favor of "beryllium."

The metal itself wasn't isolated until 1828, the year before Vauquelin died. It was produced in a very impure form by Friedrich Wöhler in Germany and Antoine-Alexandre-Brutus Bussy in France. It took another seventy years to find affordable ways to produce highly pure metal in sizeable amounts.

What's in a Name? Beryllium or Glucinum?

Vauquelin was pressured by some French chemists to name his new element "glucine," after its sweet-tasting salts, and give it the symbol "Gl." He reluctantly went along, but some objected that the name resembled "glycine," an amino acid. Others noted that yttrium (Y) also had sweet salts. Later, when Vauquelin's discovery was shown to include oxygen, those favoring "glucine" (mostly in France) renamed the element "glucinum," or "glucinium," while "beryllium" was generally used elsewhere, including in the United States and Germany. In 1949, the group with authority over element names, the International Union of Pure and Applied Chemistry, ruled that only beryllium should be used.

Identifying Minerals

	Talc	Fluorite	Quartz	Beryl	Corundum
Color	white shades	white, yellow, green, red, blue	colorless, white, gray, black, pink, purple	colorless, green, blue, yellow, pink	blue, red, yellow, gray
Luster	glassy, pearly	glassy	glassy	glassy, resinous (like dried glue)	blue, red, yellow, gray glassy
Crystal Shape	indefinite	cubic	six-sided prisms, six-sided pyramid caps	six-sided prisms, six-sided flat tops	six-sided barrels, taper to pyramid caps
Hardness	1	4	7	7.5–8	9
Streak	white	white	white	none	none

Minerals provide the raw material for the metals and chemicals we use. Many minerals can be identified by noting a few simple features. Most minerals occur in more than one color. Luster is the appearance of light reflected from the mineral's surface. Streak is the color of the powder left on a streak plate when the mineral is scraped across it.

Beryllium and the Periodic Table

All ordinary matter in the universe is made of elements and combinations of elements. There are 111 elements known today, and a few more waiting in the wings to be verified. Of the 111, 92 are found on Earth, 3 are found only in stars, and 16 are artificial. Although each element is unique, chemists over the ages have noticed that some behave like others, especially in the way they combine with other elements.

There have been many attempts to organize the elements according to chemical similarities. This is done today using a chart called the periodic table, which is based on the work of the Russian chemist Dmitry Mendeleyev. He introduced his first periodic table in 1869, which listed the sixty-three elements known at that time in order of increasing weight. He broke the element list into several segments and displayed them so that chemically similar elements lined up. This arrangement was "periodic" because it highlighted chemical behaviors that repeated in regular periods. Although his original version arranged the elements in vertical columns, with the lightest ones at the top and heaviest ones at the bottom, he eventually settled on a design that used horizontal rows, with element weights increasing from left to right.

Although the periodic table provides us with a useful summary of element properties, it was also once a powerful tool for predicting the existence of elements. Most famously, in 1871, Mendeleyev predicted the existence of three elements and their basic properties. Within fifteen years, elements were discovered (gallium [Ga], scandium [Sc], and germanium [Ge]) that closely matched his predictions. He also revised the weights of beryllium and indium (In), correctly, to line them up with groups of chemically similar elements.

Today, the periodic table arranges the elements in a grid of rows and columns according to their properties. Each column, called a group, contains elements with similar chemical behaviors. The rows, called periods, list elements according to the atomic number, which gives the number of protons in the nucleus. (See the periodic table on pages 42–43.)

Atomic Weights

Mendeleyev was especially successful because he had accurate atomic weights. He owed this to Italian chemist Stanislao Cannizzaro. In 1860, Cannizzaro cleared up the decades-long confusion about atoms and molecules, by realizing that molecules can contain more than one atom. Confusing molecules and atoms can lead to overestimates of atomic weights. It would be like weighing a carton containing a dozen eggs, but thinking it contained only one egg. Cannizzaro relied on work of fellow citizen Amedeo Avogadro from fifty years earlier that had been poorly understood at the time, and consequently ignored.

Each entry on the periodic table gives an element's name and chemical symbol, as well as the atomic number and atomic weight. The atomic weight is the average mass of the element's isotopes, reflecting the abundance of each isotope. The isotopes of an element all have the same number of protons, but differ in the number of neutrons. The atomic weight is given in terms of atomic mass units. An atomic mass unit (amu) is one-twelfth the mass of a carbon atom and is very close to the mass of a proton or one atom of hydrogen.

Beryllium's atomic number is 4, and it heads the alkaline earth metal group. Its chemical symbol is Be. Also in this group are magnesium (Mg), calcium (Ca), strontium (Sr), barium (Ba), and radium (Ra). They are named after their oxides, the alkaline earths, which, except for radium, were known as beryllia, magnesia, lime, strontia, and baryta. The oxides were actually thought to be elements until the early nineteenth century because it was so hard to break the compounds into elements. They were named alkaline earths because of their intermediate nature between the alkalis (oxides of the alkali metals) and the rare earths (oxides of rare earth metals).

Chapter Two
The Element Beryllium

An element is made up of only one kind of atom and can't be broken down any further without losing what makes it unique. Inside the atom are smaller pieces called subatomic particles.

Subatomic Details

Beryllium atoms are so small that 242 million of them laid side by side would span only an inch (2.54 centimeters). The pieces of an atom are even smaller. Although many subatomic particles are known, only three of them—protons, neutrons, and electrons—are important for understanding the basics of chemistry and physics.

At the heart of the atom is a nucleus packed with protons and neutrons. Protons have a positive electric charge and are slightly lighter than neutrons, which carry no charge.

Electrons are more than 1,800 times lighter than protons but carry a negative charge that is as strong as the proton's positive charge. In a neutral atom—one that has no net charge—there are as many electrons as protons. The electrons are found far from the nucleus, moving in complex ways. Imagining that electrons are like planets moving around a sun in definite orbits gives you a basic mental picture of this situation. However, the electrons can move only in orbits, or "shells," corresponding to certain energies. The number of electrons allowed in a shell is limited. When a shell fills up, electrons move to the next one. The first shell is nearest to the nucleus and can hold two electrons, while the next shell can hold eight electrons.

Elements and Atomic Structure

The most important property of an element that makes it different from any other element is the number of protons in its nucleus. This property controls the typical number of electrons and neutrons the element can have. The arrangement of the electrons, in turn, affects how elements combine with each other.

Depending on the proton count, an element is categorized as a metal, metalloid, or nonmetal. The difference of a few protons can lead to dramatically different properties. Metals tend to lose their electrons, while nonmetals tend to pick them up. Also, all metals (except mercury [Hg]) are solid at room temperature, good conductors of heat and electricity, and are generally easy to hammer into shapes or drawn out into wires. Nonmetals have properties opposite those of metals. They are poor conductors of electricity and heat, brittle, and not easily worked into shapes or wires. Metalloids have both metallic and nonmetallic properties.

Beryllium, a strong metal, has four protons. The element with five protons is boron (B), a metalloid. The group's most economically important properties are its poor electrical conduction at low temperatures and its moderate conduction at higher temperatures. This behavior means the conductivity can be controlled, and it is what makes the metalloid silicon so valuable in the semiconductor and computer chip industries.

The element with one fewer proton than beryllium is lithium (Li), an alkali metal. As another metal, lithium conducts electricity and heat well. But while beryllium has great strength, lithium can be cut with a knife. It also ignites and burns in air when dropped onto water, though not quite as violently as sodium (Na) and potassium (K), two other alkali metals.

An atom of beryllium has four protons and five neutrons in its nucleus. Four electrons orbit the nucleus in two shells. Electrons in the outermost shell are valence electrons.

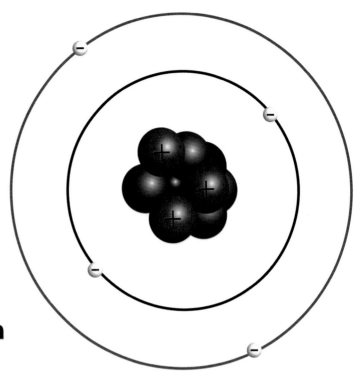

Another Visit with the Periodic Table

Mendeleyev's periodic table listed elements in order of atomic weight. Although this arrangement allowed him to find patterns of chemical similarity, it failed in a few cases. For example, when the gas argon (Ar) was discovered in 1894, its atomic weight of 39.95 placed it after potassium, with an atomic weight of 39.10, and broke the pattern that grouped potassium with sodium and lithium. But the pattern was restored twenty years later when Henry Moseley showed that the atomic number gives the positive charge of the nucleus and that listing elements in order of atomic number instead of atomic weight keeps the periodic table in complete agreement with chemical properties.

Like Mendeleyev before him, Moseley successfully predicted three elements using the periodic table, but based on atomic number, not atomic weight. These had the atomic numbers 43 (technetium [Tc], discovered in 1937), 61 (promethium [Pm], discovered in 1945), and 75 (rhenium [Re], discovered in 1925).

Beryllium Snapshot

Chemical Symbol:	Be
Classification:	Alkaline earth metal
Properties:	High thermal conductivity, permeable to X-rays, high stiffness
Discovered By:	Nicolas-Louis Vauquelin in 1787
Atomic Number:	4
Atomic Weight:	9.012 atomic mass units (amu)
Protons:	4
Electrons:	4
Neutrons:	5
Density at 68°F (20°C):	1.85 grams per cubic centimeter (g/cm³)
Melting Point:	2,349°F; 1,287°C
Boiling Point:	4,476°F; 2,469°C
Commonly Found:	Granitic rocks, pegmatites, streambeds

The reason Mendeleyev's grouping by atomic weight worked as well as it did is because the elements usually have as many neutrons as protons, but not always. In a very few cases, this reverses the ordering, as it did for argon, with eighteen protons and twenty-two neutrons, and potassium, with nineteen protons and twenty neutrons.

Columns and Valence, Periods, and Shells

The periodic table's groups, or columns, contain elements that combine in similar ways with other elements. For example, one atom of any alkaline earth element, including beryllium, combines with one atom of oxygen (O).

These chemical properties depend on the valence electrons, which are the electrons in the outermost energy level of an atom. Valence electrons can be lost, gained, or shared. The alkaline earth metals have two valence electrons. Elements following this simple rule are labeled with the valence in Roman numerals and an "A." These are called main group elements. Beryllium's group is IIA.

Atomic Number

Before 1913, nobody was sure what the atomic number meant. At the time, it was only the place number in Mendeleyev's periodic table listing elements in order of weight. In 1913, English physicist Henry G. J. Moseley showed that the atomic number gave the positive charge of the nucleus. Showing that the atomic number is the proton count had to wait until the 1932 discovery of neutrons in the nucleus. Until then, some scientists thought there could be neutral combinations of protons and electrons.

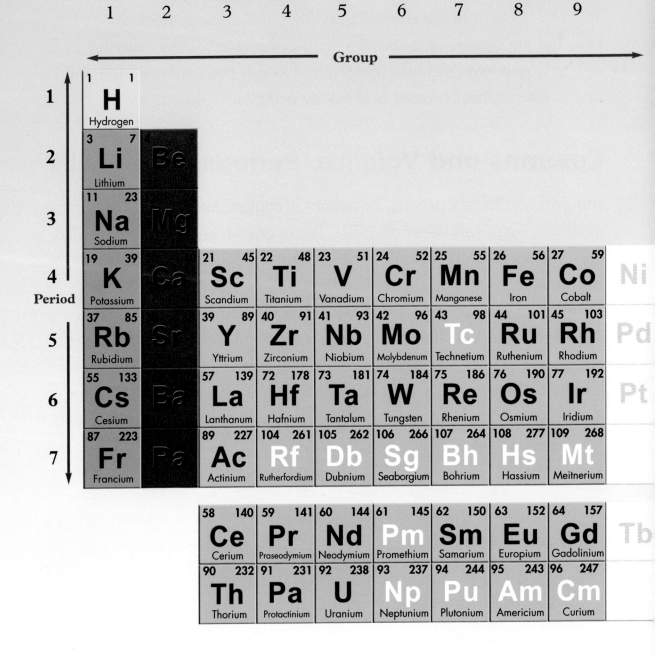

	IA 1	IIA 2	IIIB 3	IVB 4	VB 5	VIB 6	VIIB 7	VIIIB 8	VIIIB 9

Group

Period										
1	1 1 **H** Hydrogen									
2	3 7 **Li** Lithium	7 **Be**								
3	11 23 **Na** Sodium	**Mg**								
4	19 39 **K** Potassium	**Ca**	21 45 **Sc** Scandium	22 48 **Ti** Titanium	23 51 **V** Vanadium	24 52 **Cr** Chromium	25 55 **Mn** Manganese	26 56 **Fe** Iron	27 59 **Co** Cobalt	Ni
5	37 85 **Rb** Rubidium	**Sr**	39 89 **Y** Yttrium	40 91 **Zr** Zirconium	41 93 **Nb** Niobium	42 96 **Mo** Molybdenum	43 98 **Tc** Technetium	44 101 **Ru** Ruthenium	45 103 **Rh** Rhodium	Pd
6	55 133 **Cs** Cesium	**Ba**	57 139 **La** Lanthanum	72 178 **Hf** Hafnium	73 181 **Ta** Tantalum	74 184 **W** Tungsten	75 186 **Re** Rhenium	76 190 **Os** Osmium	77 192 **Ir** Iridium	Pt
7	87 223 **Fr** Francium	**Ra**	89 227 **Ac** Actinium	104 261 **Rf** Rutherfordium	105 262 **Db** Dubnium	106 266 **Sg** Seaborgium	107 264 **Bh** Bohrium	108 277 **Hs** Hassium	109 268 **Mt** Meitnerium	

58 140 **Ce** Cerium	59 141 **Pr** Praseodymium	60 144 **Nd** Neodymium	61 145 **Pm** Promethium	62 150 **Sm** Samarium	63 152 **Eu** Europium	64 157 **Gd** Gadolinium	Tb
90 232 **Th** Thorium	91 231 **Pa** Protactinium	92 238 **U** Uranium	93 237 **Np** Neptunium	94 244 **Pu** Plutonium	95 243 **Am** Americium	96 247 **Cm** Curium	

Beryllium is on the left side of the periodic table, making it a metal. It is a good conductor of heat and electricity. Beryllium is in group IIA, known as the alkaline earth metal group.

Mendeleyev knew about valence, and this was how he grouped the elements. But he didn't know about electrons, which were discovered thirty years after he published his first table.

Mendeleyev's periods—the rows of the table—correspond to different shells. Beryllium is in period 2, so it has two electron shells. Because it is in group 2, we know there are two electrons in the outermost shell.

Some elements follow more complicated rules for electrons, progressively filling other kinds of electron sub-shells (called *d* and *f* orbitals), which allows them to combine in more than one way with oxygen. Elements with electrons in one or more of these orbitals are transition metals (in columns labeled with "B"), and the lanthanide (La) and actinide (Ac) series (in the bottommost rows).

Group Number

As more was learned about the quantum mechanical behavior of electrons in atoms, the correspondence between periodic table group numbers and valence electrons became clear. It also became more complicated because of the "inner orbital" *d*, which begins to fill up in period 4 of the table, starting with scandium. The International Union of Pure and Applied Chemistry decided to label the columns to bring the periodic table into accord with the modern understanding of electron structure in atoms. Rather than having "A" and "B" columns with Roman numerals, the column labels increase from 1 through 18, specifying the total number of electrons added to the atom since the last noble gas element.

A Closer Look at Groups

We saw that boron, the next element after beryllium, is a metalloid. In the periodic table, it is at the top of a stair-stepping band that ends at element 84, polonium. The elements within the band are metalloids. Those to the left are metals. All the metals are solid at normal conditions, except for mercury, and are shiny, good conductors of heat and electricity, and for the most part ductile and malleable.

Elements to the right of the metalloid band are nonmetals, which include all the gas elements. The solid nonmetals are brittle, poor conductors of heat and electricity, not malleable or ductile, and tend to gain electrons in chemical reactions.

Chapter Three
Beryllium's Properties

All matter has characteristic properties that help identify it. Physical properties can be detected without transforming one substance into another. For example, melting a piece of beryllium changes it from a solid to a liquid, but it is still beryllium, so melting temperature is a physical property. Other properties may be chemical. Chemical properties involve the way a substance changes (or doesn't change) when combined with other substances. One atom of beryllium combines with one atom of oxygen, so it has two valence electrons, which is a chemical property. Finally, beryllium has nuclear properties—features of its nucleus—that set it apart from other elements.

Physical Properties

Beryllium is a light metal. It is a solid under normal conditions, it conducts electricity and heat well, and a chunk of it is lighter than a same-sized chunk of steel, for example. In fact, beryllium is four times lighter than steel and 30 percent lighter than aluminum, another light metal.

Density helps us measure lightness by telling us how much mass we can fit into a standard volume. Density could be measured as slugs per soda bottle, but the standard in chemistry is grams per cubic centimeter. Beryllium's density at room temperature is 1.85 g/cm^3.

Mohs' Scale

Hardness Rating Examples

1 Talc
2 Gypsum (rock salt, fingernail)
3 Calcite (copper [Cu])
4 Fluorite (and iron [Fe])
5 Apatite (and cobalt [Co])
5.5 Beryllium
6 Orthoclase (and rhodium [Rh], silicon, tungsten [W])
7 Quartz
8 Topaz (and chromium [Cr], hardened steel)
9 Corundum (sapphire)
10 Diamond

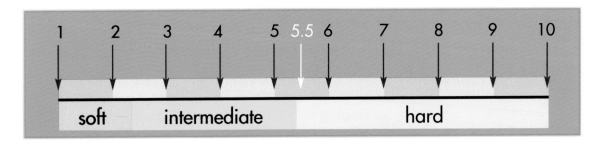

Hardness is the ability of a substance to resist scratching. German mineralogist Friedrich Mohs developed a simple hardness scale almost 200 years ago that is still used. It ranks a mineral's hardness by comparing it with the hardness of ten index minerals.

Not only is beryllium a good heat conductor, it also has high heat capacity, meaning its temperature doesn't change much when heat is applied. Moreover, its 2,349°F (1,287°C) melting temperature is very high for a light metal, almost 1,112°F (600°C) higher than aluminum.

Beryllium has exceptional stiffness, so things made of beryllium tend to keep their shapes when squeezed or twisted. Relative to density, beryllium is six times stiffer than steel and aluminum. It has moderate hardness, with a rank of 5.5 on the Mohs' 10-point hardness scale. The Mohs' scale was invented to rate mineral hardness, and it ranges from 1 for talc to 10 for diamond. Beryllium is harder than iron but not as hard as glass. Optically, polished beryllium is a good reflector of infrared light but is transparent to X-rays. For all these outstanding properties, beryllium has poor malleability and ductility, meaning it is hard to squeeze into shapes and stretch into wires.

Chemical Properties

Beryllium is reactive at high temperatures, where it readily forms compounds with nonmetals such as oxygen and chlorine (Cl). The atoms in the compounds lock tightly in place by sharing beryllium's two valence electrons. This kind of link is called a covalent bond.

The other members of the alkaline earth metal group usually form ionic bonds, which involve the attraction between positive and negative ions. These group members give up their valence electrons to gain a positive charge of 2 and favor links with elements and compounds that have a negative charge of 2.

Beryllium's covalent bonding is due to the atom's small size compared with the other group members. Only two electrons lie between beryllium's nucleus and two valence electrons, while ten electrons are between the nucleus and two valence electrons of magnesium, the next element in the group. Beryllium's valence electrons are therefore closer to the nucleus than magnesium's and also encounter less interference from inner electrons. These two effects—closeness to the nucleus and less inner-electron interference—help beryllium keep a tight hold on its valence electrons, so it never exists as a free ion with a positive 2 charge. Atom sizes also account for the chemical similarity of beryllium and aluminum, an element with three valence electrons.

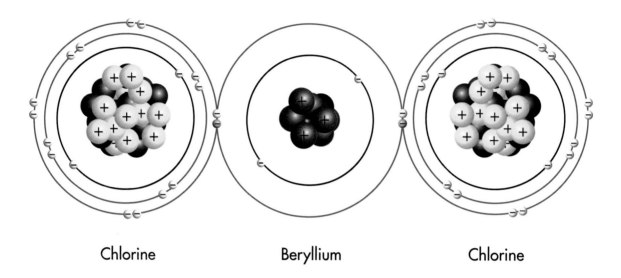

Chlorine   Beryllium   Chlorine

One beryllium atom combines with two chlorine atoms using strong covalent bonds. The covalent bond between two atoms pairs one valence electron from each atom. The attraction of each nucleus on the electron pairs locks the atoms tightly in place.

Nuclear Properties

Beryllium has several isotopes, each with a unique number of neutrons. Some isotopes are radioactive, which means the nucleus changes all by itself, with no outside help. Most of the time, this change transforms the nucleus into another element. This is called radioactive decay because given enough time, all of the radioactive isotope will be gone. A measure of the speed of decay is the half-life, which tells how long it takes for half of the isotope to change.

All of beryllium's isotopes are radioactive except beryllium-9, which has five neutrons. Beryllium-10, with six neutrons, has a half-life of about 1.5 million years and decays into boron-10, which is stable. Beryllium-7 has a fifty-three-day half-life and decays into lithium-7,

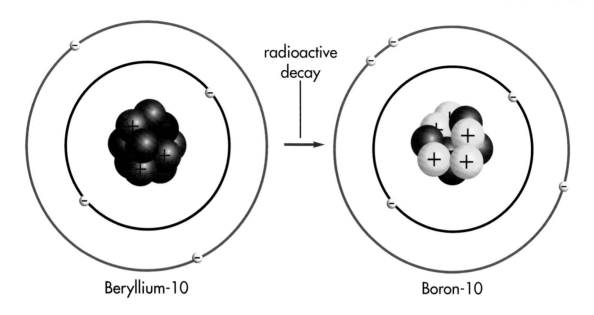

Beryllium-10   radioactive decay   Boron-10

Radioactive decay usually turns one element into another. Radioactive beryllium-10 decays into stable (non-radioactive) boron-10. The nucleus of beryllium-10 has four protons and six neutrons. When decay occurs, one of the six neutrons splits into a proton and an electron. The electron is ejected from the nucleus, leaving behind the five protons and five neutrons of boron-10.

while beryllium-8, with four neutrons, breaks into two helium (He) atoms almost the instant it is created. All beryllium isotopes, including beryllium-9, are created when cosmic rays—particles and atoms ejected from exploding stars—shatter heavier elements such as oxygen and nitrogen (N). This process, called spallation, occurs in the depths of interstellar space, as well as closer to home in the atmosphere, water, and rocks. All boron and most lithium are also created this way. The other elements were created when the universe began (hydrogen [H] and most helium) or in stars (the rest).

Beryllium is a strong neutron reflector and is able to slow down fast neutrons. These properties are used in nuclear fission applications, such as power generation and weapons.

Chapter Four
Where Can Beryllium Be Found?

Although beryllium is rare, it is an important part of more than 100 minerals. It is found in rocks, soil, water, food, and the air. Most of Earth's beryllium has been here since the planet formed 4.5 billion years ago, locked in common rock-forming minerals and not in beryllium-rich minerals. Concentrations of beryllium occur where hot water dissolves rocks, such as volcanic areas and where tectonic plates collide. In addition, beryllium is created in the upper atmosphere and in rocks at or near the surface by cosmic rays shattering oxygen and nitrogen.

Mining and Processing Beryllium

The minerals bertrandite and beryl are the only commercially important sources of beryllium. At present, the bertrandite being mined is concentrated in western Utah, while the beryl is from South America, China, and Kazakhstan. Gemstone beryl, which includes emerald and aquamarine, comes mostly from Brazil. Large deposits of another beryllium mineral, chrysoberyl, have been found in a remote part of Alaska, but they are not being mined.

Beryl contains about 4 percent beryllium by weight, and bertrandite has less than 1 percent. While bertrandite is lower in content than beryl,

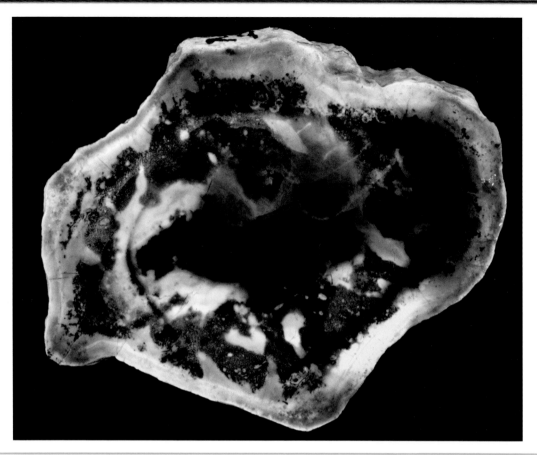

Beryllium ores are sometimes associated with the intensely purple fluorine mineral fluorite, shown in this mineralized nodule. The minerals in the nodule's interior are purple fluorite and white opal and quartz. The outer zone is fluorite and opal. The beryllium is concentrated in the fluorite. The nodule starts out as a limestone-like rock and is altered by hot fluorine-rich volcanic water. Over time, fluorine and silicon replace the minerals in the original rock.

it can be mined from shallow pits and is more easily processed. Also, suitable beryl crystals must be laboriously sorted by hand.

Although the details differ depending on the ore, both types of source rocks are crushed, leached with sulfuric acid, and heat-treated to dissolve the beryllium. Further processing produces beryllium hydroxide, $Be(OH)_2$, which can be refined into purified metal, alloys, and oxides.

Refining the beryllium hydroxide to extract the metal involves a kind of chemical bucket brigade that is common in the chemical industry. The beryllium is passed from one compound to another until it can be separated with a small amount of effort. This procedure takes only a few steps for some elements, but beryllium isn't one of them! In a seven-step process that includes a lot of heat and compounds of fluorine (F), magnesium, calcium, and lead, a fairly pure metal comes out the other end.

This last step isn't the end of the road, though. Beryllium can't be stretched or extruded (squeezed) very easily. Instead, it is hammered and chipped into fine powders that can be pressed into shapes and baked near the melting temperature to bind the particles together.

In addition to processing ore to get new beryllium, some demand is met through recycling and reuse. Most of this is scrap left over after making parts from beryllium. Reclaimed metal, such as connectors on computer boards, supplies the rest, although more than 90 percent of this is not retrieved.

At times, commercial U.S. demands have been met through purchases from the government's stockpiles and through imports of lower-grade material from Kazakhstan for alloys.

Beryllium Metal Uses

Despite beryllium's high cost, it is often used because nothing else will get the job done. The combination of lightness, stiffness, and thermal properties make it ideal for satellite structures, the National Aeronautics and Space Administration's space shuttle's window frame, and gyroscopes in guidance systems. These parts keep their shapes and don't flex very much under a wide range of temperatures and stresses.

These same properties, together with beryllium's ability to reflect infrared energy, make it ideal for space-based infrared telescopes. Satellite missions dating back to the 1970s have used the metal for telescope mirrors. The James Webb Space Telescope, scheduled for launch sometime after 2010, will have a 21-foot-wide (6.4-meter-wide) beryllium mirror. The mirror consists of eighteen polished six-sided beryllium segments coated with gold for even better reflection. The mirror will be six times bigger than

The mirror of the James Webb Space Telescope consists of eighteen six-sided petals of pure beryllium metal like the one shown here. The metal is so strong and rigid that most of a segment's underside can be replaced by a lightweight honeycombed structure. The other side of the mirror will be coated with a thin film of gold to increase the reflection of infrared light.

the Hubble Space Telescope's glass mirror, but at 830 pounds (376 kilograms), it weighs less than half the Hubble's 1,840 pounds (835 kg).

Other uses of beryllium exploit its transparency to X-rays. The metal is used as both a window in X-ray generating tubes to let the penetrating radiation out, and as a lens material to focus X-rays.

Neutrons and Isotopes

Beryllium's emission of neutrons when it is bombarded by subatomic particles has been used to treat cancer, detonate nuclear weapons, and probe subatomic secrets. You have already seen how this property helped discover the neutron itself. Today, particle accelerators hurl particles at beryllium targets to generate neutrons.

Nuclear power plants can use beryllium to cut down on lost power. The plants make power by capturing the heat generated when neutrons split uranium (U). In a "chain reaction," this also generates additional neutrons that sustain uranium splitting. Beryllium is used in some reactors to reflect stray neutrons back into the uranium core to keep them from leaking out. It also slows fast neutrons, which don't do a very good job of splitting uranium.

A different nuclear feature of beryllium is very useful for geological detective work. The accumulation of the radioactive isotope beryllium-10 can tell us when some things were buried and when other things were

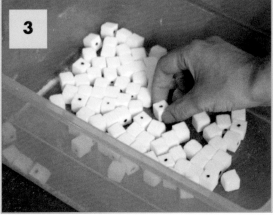

Radioactive decay can be simulated by a simple experiment. Here, sugar cubes marked with black dots represent radioactive atoms (1). After rotating a box of the cubes to roll them (2), the black dot of some cubes will be face up (3), indicating radioactive decay. A graph of how many cubes remain plotted against the number of times they have been rolled looks much like a radioactive decay curve.

uncovered. In either case, geologists rely on the facts that a rock accumulates beryllium-10 when cosmic rays strike the oxygen in its minerals and that if the rock is buried more than a few meters deep, the cosmic rays can't reach it.

As an example of this geologic detective work, a recent study using beryllium-10 measurements determined that glaciers from the Laurentide ice sheet first bulldozed North America about 2.4 million years ago. The study assumed that before they were buried by the ice sheet, the rocks had a steady amount of the isotope in them, meaning that cosmic rays generated as much new isotope as the rocks lost through radioactive decay. Once buried, cosmic rays could no longer replenish the isotope, which decayed away. Using these assumptions, the scientists were able to figure out how long the rocks had been buried based on the amount of the beryllium-10 in them today.

Chapter Five
Beryllium Compounds and Alloys

Beryllium combines with metals and nonmetals to form two very different kinds of substances. The metal combinations are called alloys and have metallic properties such as malleability and good electrical conductivity. The nonmetal combinations are called compounds, and generally don't have metallic properties.

Metals owe their properties to their loosely held electrons. In fact, chemists often describe metals as consisting of metal ions floating in a sea of electrons. The attractions between the positively charged metal ions and the "sea" of easy-moving electrons hold them all together in a network of metallic bonds.

The compounds that beryllium forms with nonmetals are held together by covalent bonds. In covalent bonding, valence electrons are shared between atoms. The attraction of each atom on the shared electrons is what binds the atoms together.

A key feature of compounds is that they always form in the same proportions. Beryllium oxide (BeO) always pairs one beryllium atom with one oxygen atom. Alloys, on the other hand, can mix in arbitrary proportions.

Beryllium's use of the covalent bond is unusual. Metals and nonmetals usually combine with ionic bonds. The metal loses valence electrons that the nonmetal gains, and the atoms become oppositely charged ions that attract each other. Alternating metal and nonmetal ions can network

Beryllium (yellow) combines with oxygen (red) to form beryllium oxide. Usually, a metal (beryllium) and nonmetal (oxygen) combine with ionic bonds. However, beryllium's valence electrons orbit so close to the protons in the nucleus that the stronger pull forms a covalent bond.

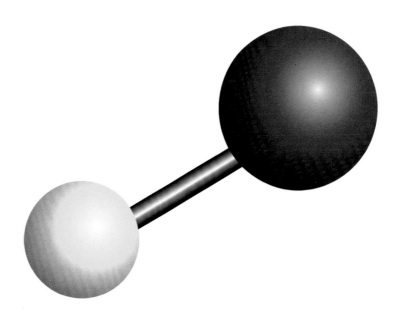

to form a crystal structure. All the members of the alkaline earth metal group except beryllium can form ionic compounds. Beryllium's small size favors covalent bonding because its two valence electrons can't break the pull from the protons in the nucleus.

Beryllium Mineral Compounds

In addition to providing sources of raw beryllium, some of the minerals the element forms are used as gemstones. Emerald and aquamarine are glamorous beryllium compounds, especially when they are cut to emphasize the symmetry of their crystal structure. They differ from the lowly rock-forming mineral beryl only in their colors and relative rarity. Beryl can be green, blue, yellow, pink, or red.

Chrysoberyl, another colorful beryllium mineral, can be yellow, green, or yellow-green. Unusual varieties of it include cat's eye and the color-changing alexandrite, which looks green in sunlight and red under lightbulb illumination.

Crystals of the blue-green gemstone aquamarine shown here are a type of beryl, which contains the element beryllium. Until plentiful deposits of bertrandite were found in Utah, most beryllium and beryllium compounds were derived from processed beryl.

Cheating Nature

Sapphires are forms of corundum (aluminum oxide) other than the red ruby, and come in yellow, green, and purple varieties. Some people have been using beryllium to make common corundum crystals look like rare (and expensive) sapphires, and pass off the result as natural. This deceitful practice diffuses tiny beryllium atoms from the mineral chrysoberyl into the corundum crystals by baking them together at high temperatures. The result is that an inexpensive pink sapphire can be gussied up to look like the prized and rare orange-pink sapphire called padparadscha. Many tens of thousands of the "fake" gems had been certified and purchased by 2003, when the fraud was uncovered. Today, a range of tests is available to detect when beryllium has been used to "tune" a sapphire's color.

Beryllium Alloys

About 70 percent of beryllium's use is in alloys, and most of the alloys involve copper (Cu). Beryllium is such a small atom that it occupies the spaces between the host metal atoms. Even though it is usually present in amounts less than 10 percent by weight, its influence on the alloy can be strong.

Alloys have been made with copper, nickel (Ni), and aluminum. Including small amounts of cobalt (Co), nickel, silicon, or lead fine-tune the alloy's performance. Beryllium strengthens copper and nickel, and improves aluminum's oxidation resistance, as well as its ability to be cast and worked.

Copper-beryllium alloys are used for current-carrying components in electronics and computers that need to stand up to repeated wear and flexing. These same properties make the alloys ideal for steering-wheel connecting springs and springs in wind-up music boxes, and for clips on pens. The combination of hardness and nonsparking gives tools made from these alloys an edge over steel in firefighting and hazardous material settings. These alloys are used in molds for casting fine details and textures like wood grain into plastic parts because of their wear-resistance and high thermal conductivity. They are also used in radiation shielding on cell phones.

Beryllium Ceramics

The most important beryllium compound is beryllium oxide, also called beryllia, which is used in ceramic form. Ceramics are made by firing beryllia parts, much like a clay pot would be fired in a kiln.

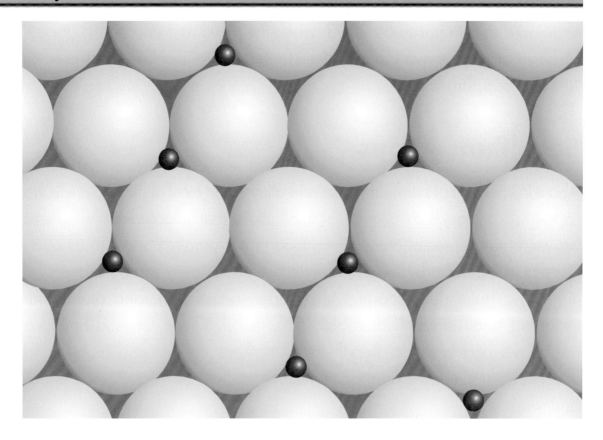

In an alloy of copper and beryllium, beryllium atoms (red) occupy the spaces between the much larger copper atoms (green). Even though present in small amounts, the effect of the beryllium on alloy strength is large.

The material withstands extreme temperatures and dissipates heat faster than any other ceramic. It is also a good electrical insulator, which means it conducts electricity poorly. These properties provide great benefits in electronic applications that generate potentially damaging heat, such as high-speed computers and automobile electronic ignition systems. Mounting components on a base of beryllia ceramic allows the heat to be quickly whisked away.

These properties facilitate small and long-lasting ion lasers, which are used in printing, eye surgery, and DNA analysis. The ceramic laser body dissipates the heat generated when high currents arc through the laser's gas. The laser body lasts longer because of the ceramic's high hardness, strength, and corrosion resistance.

Argon lasers, like the one shown here in the background, are used in surgery and industrial applications. They generate a great deal of heat and must often be run for long times. At the heart of many argon lasers is a beryllium oxide (BeO) ceramic plasma tube. The BeO plasma tube can withstand the heat.

Deadly Dust

Breathing fine beryllium particles can cause a debilitating lung disorder called chronic beryllium disease. The symptoms include persistent coughing, difficulty breathing with physical exertion, fatigue, chest and joint pain, weight loss, and fevers. It strikes up to 10 percent of those who have inhaled beryllium dust or fumes, and it can take anywhere from a few months to thirty years to develop. Occupational exposure often occurs in mining and extraction, as well as in processing beryllium alloys. Beryllium materials, in solid form and finished products, present no special health risks.

This dark side of beryllium began to emerge in the 1930s when a cluster of workers in the fluorescent lamp industry reported breathing problems and skin lesions. A 1946 study traced the symptoms to beryllium compounds used in fluorescent bulbs. By 1949, many more cases had been diagnosed, and the lamp industry abandoned the use of beryllium. The disease also afflicted some who helped create the first atomic bombs, which had several beryllium components.

s rare and potentially dangerous as beryllium can be, it crops up in surprisingly ordinary places. The element and its compounds find applications where exceptional heat dissipation or springiness is needed. The element also plays a crucial role in the creation of most elements and in making life as we know it possible.

Bar-code Readers

Bar-code scanning is used in many places for rapid and accurate information handling. Whether the bar-code reader is the stationary scanner at a grocery's checkout counter or a handheld scanner that a parcel delivery service uses to track a package, chances are they have moving parts made from beryllium-copper alloy. The alloy allows faster operation at lower power because of its lightness and ability to quickly spring back to shape when flexed, a property called stiffness. The alloy's strength also helps protect handheld readers from shock and misalignment if they are dropped.

High-Fidelity Speakers

Music purists have found that speakers with diaphragms made from beryllium metal have an unequaled ability to match the original sound.

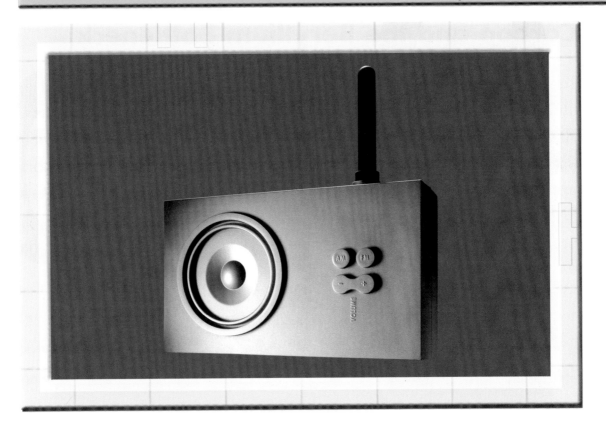

Sound is made when air is rapidly squeezed and stretched. A hi-fi loudspeaker does this by pushing and pulling a cone-shaped diaphragm. Beryllium diaphragms, such as the one in this wireless music player, are exceptionally good at producing high tones because of the metal's stiffness.

The secret lies in how beryllium's high stiffness affects the response when the diaphragm is vibrated to make sound. Typically, a speaker overreacts at high tones, where the vibrating part resonates. At a resonant frequency, all parts of the diaphragm move in synchrony, causing unusually loud sound. It's like a trampoline where several people are able to synchronize their jumps to produce a huge bounce. Beryllium's high stiffness, together with its light weight, move the resonances to frequencies beyond the range of hearing. The resonance is moved to higher frequencies because the metal flexes back into shape more quickly than less stiff speaker material such as aluminum.

Carbon and the Beryllium Bottleneck

The isotope beryllium-8 has a special place in the universe. Without it, carbon-12 wouldn't exist in the amounts that make life possible on Earth. However, the amazing fact is that beryllium-8 is extremely delicate, splitting into two helium-4 atoms almost the instant it is created inside stars from two other helium-4 atoms. In the 1940s and 1950s, when scientists first started solving the puzzle of how elements are created, this quick split posed a theoretical problem called the beryllium bottleneck. It seemed the

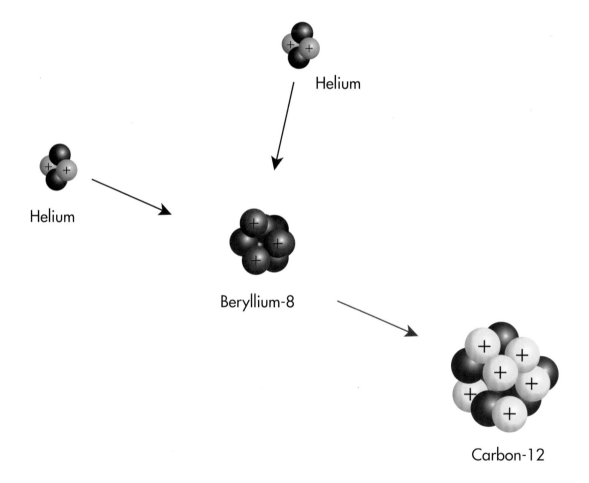

Helium

Helium

Beryllium-8

Carbon-12

Two helium atoms combine in the extreme temperatures of red suns to create an atom of short-lived beryllium-8. The beryllium immediately decays into an atom of carbon-12.

isotope wasn't around long enough to act as a building block for the heavier elements in a process called nucleosynthesis. The process explains how to build many of the heavier elements starting from carbon-12 and even explains how to build carbon-12 from beryllium-8 and helium-4 atoms. However, based on what was known at the time, the universe should have far less carbon. It was as though every time a brick was laid, it was whisked away before another could be put on top of it. The solution was found when scientists realized that in hot stars called red giants, the beryllium and helium collisions have a much better chance of converting into carbon before the beryllium splits.

Water Bottles and Beryllium

The plastic bottle that holds the water you drink might have been made with the help of a beryllium-copper alloy. In one manufacturing process, molten plastic is inflated like a balloon into a bottle-shaped mold made of the alloy. The alloy's rapid heat transfer allows the whole bottle to cool evenly and quickly. This fast heat transfer avoids warping and speeds up bottle production.

A Special Element

Beryllium is a rare and special element. It plays a crucial role in helping stars create the heavier elements, including the carbon that forms our life. Beryllium itself is created when cosmic rays shatter some of these heavier elements. Beryllium helped usher in the nuclear age and was a key tool in the discovery of the neutron. Its gemstone minerals emerald and alexandrite bring us delight, and its amazing alloys and ceramics enhance manufacturing, commerce, electronics, and entertainment. For such a small and precious element, beryllium has an astonishingly wide influence.

The Periodic Table of Elements

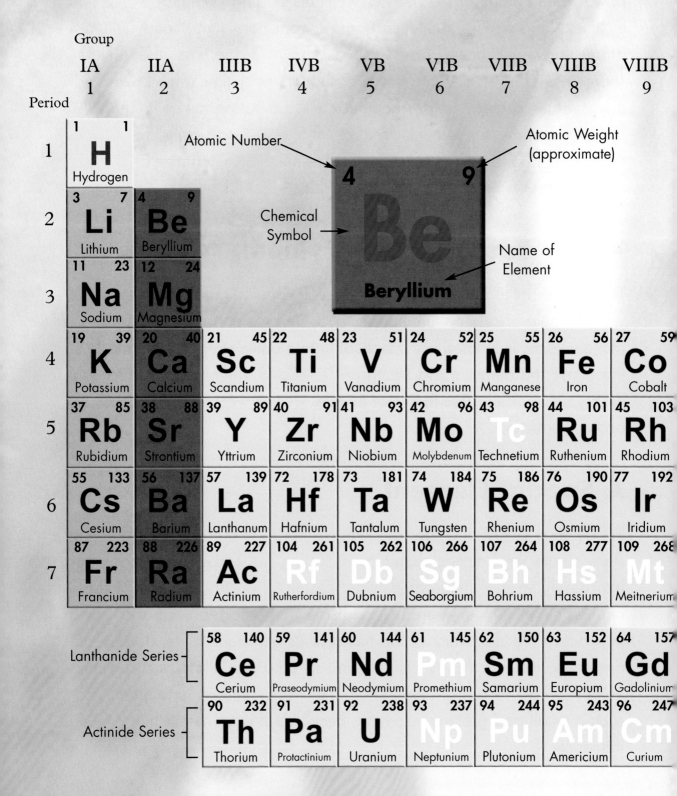

Group

IA	IIA	IIIB	IVB	VB	VIB	VIIB	VIIIB	VIIIB
1	2	3	4	5	6	7	8	9

Period

Atomic Number

Atomic Weight (approximate)

Chemical Symbol

Name of Element

4 9 Be Beryllium

1 — 1 1 H Hydrogen

2 — 3 7 Li Lithium — 4 9 Be Beryllium

3 — 11 23 Na Sodium — 12 24 Mg Magnesium

4 — 19 39 K Potassium | 20 40 Ca Calcium | 21 45 Sc Scandium | 22 48 Ti Titanium | 23 51 V Vanadium | 24 52 Cr Chromium | 25 55 Mn Manganese | 26 56 Fe Iron | 27 59 Co Cobalt

5 — 37 85 Rb Rubidium | 38 88 Sr Strontium | 39 89 Y Yttrium | 40 91 Zr Zirconium | 41 93 Nb Niobium | 42 96 Mo Molybdenum | 43 98 Tc Technetium | 44 101 Ru Ruthenium | 45 103 Rh Rhodium

6 — 55 133 Cs Cesium | 56 137 Ba Barium | 57 139 La Lanthanum | 72 178 Hf Hafnium | 73 181 Ta Tantalum | 74 184 W Tungsten | 75 186 Re Rhenium | 76 190 Os Osmium | 77 192 Ir Iridium

7 — 87 223 Fr Francium | 88 226 Ra Radium | 89 227 Ac Actinium | 104 261 Rf Rutherfordium | 105 262 Db Dubnium | 106 266 Sg Seaborgium | 107 264 Bh Bohrium | 108 277 Hs Hassium | 109 268 Mt Meitnerium

Lanthanide Series — 58 140 Ce Cerium | 59 141 Pr Praseodymium | 60 144 Nd Neodymium | 61 145 Pm Promethium | 62 150 Sm Samarium | 63 152 Eu Europium | 64 157 Gd Gadolinium

Actinide Series — 90 232 Th Thorium | 91 231 Pa Protactinium | 92 238 U Uranium | 93 237 Np Neptunium | 94 244 Pu Plutonium | 95 243 Am Americium | 96 247 Cm Curium

42

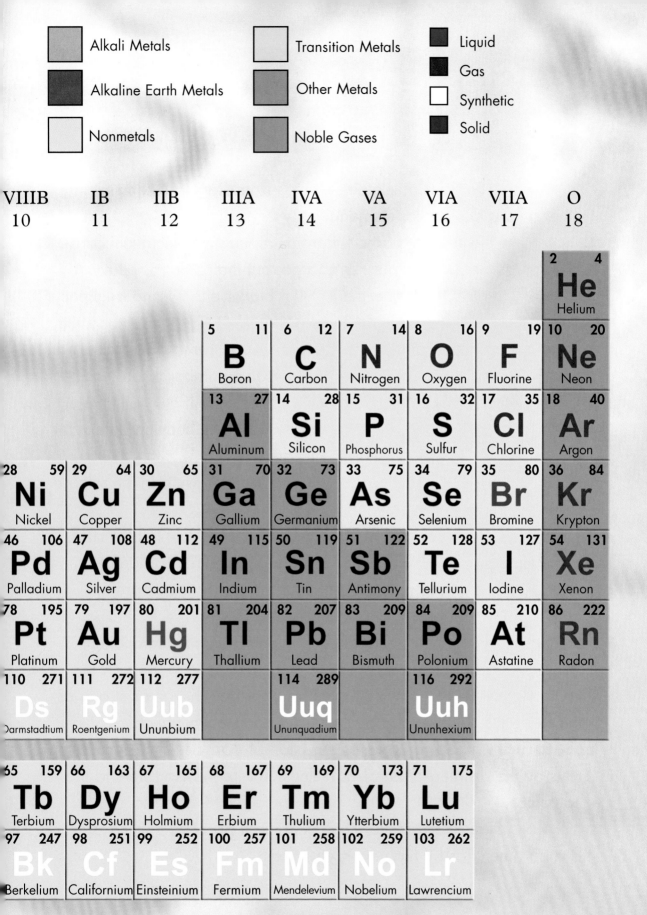

Legend

- Alkali Metals
- Alkaline Earth Metals
- Nonmetals
- Transition Metals
- Other Metals
- Noble Gases
- Liquid
- Gas
- Synthetic
- Solid

VIIIB 10	IB 11	IIB 12	IIIA 13	IVA 14	VA 15	VIA 16	VIIA 17	O 18
								2 4 **He** Helium
			5 11 **B** Boron	6 12 **C** Carbon	7 14 **N** Nitrogen	8 16 **O** Oxygen	9 19 **F** Fluorine	10 20 **Ne** Neon
			13 27 **Al** Aluminum	14 28 **Si** Silicon	15 31 **P** Phosphorus	16 32 **S** Sulfur	17 35 **Cl** Chlorine	18 40 **Ar** Argon
28 59 **Ni** Nickel	29 64 **Cu** Copper	30 65 **Zn** Zinc	31 70 **Ga** Gallium	32 73 **Ge** Germanium	33 75 **As** Arsenic	34 79 **Se** Selenium	35 80 **Br** Bromine	36 84 **Kr** Krypton
46 106 **Pd** Palladium	47 108 **Ag** Silver	48 112 **Cd** Cadmium	49 115 **In** Indium	50 119 **Sn** Tin	51 122 **Sb** Antimony	52 128 **Te** Tellurium	53 127 **I** Iodine	54 131 **Xe** Xenon
78 195 **Pt** Platinum	79 197 **Au** Gold	80 201 **Hg** Mercury	81 204 **Tl** Thallium	82 207 **Pb** Lead	83 209 **Bi** Bismuth	84 209 **Po** Polonium	85 210 **At** Astatine	86 222 **Rn** Radon
110 271 **Ds** Darmstadtium	111 272 **Rg** Roentgenium	112 277 **Uub** Ununbium		114 289 **Uuq** Ununquadium		116 292 **Uuh** Ununhexium		

65 159 **Tb** Terbium	66 163 **Dy** Dysprosium	67 165 **Ho** Holmium	68 167 **Er** Erbium	69 169 **Tm** Thulium	70 173 **Yb** Ytterbium	71 175 **Lu** Lutetium
97 247 **Bk** Berkelium	98 251 **Cf** Californium	99 252 **Es** Einsteinium	100 257 **Fm** Fermium	101 258 **Md** Mendelevium	102 259 **No** Nobelium	103 262 **Lr** Lawrencium

Glossary

cosmic rays Particles and atoms ejected from stars that can smash other atoms into smaller elements.

electron An elementary particle that has a negative electrical charge as large as the proton's positive charge, but that is not as heavy.

gamma rays Electromagnetic energy of higher energy and frequency than X-rays.

half-life The time in which half of a radioactive element disintegrates.

isotope Any of several atoms having the same number of protons but different numbers of neutrons.

neutron An uncharged elementary particle slightly heavier than a proton and found primarily in atomic nuclei.

nuclear fission The splitting of a nucleus, which results in the release of large amounts of energy.

nucleosynthesis Production of heavier elements from lighter ones.

nucleus (plural: nuclei) The positively charged central portion of an atom containing nearly all the atomic mass, consisting of protons and, usually, neutrons.

ore A naturally occurring mineral containing a valuable component that is mined and worked.

proton A particle that carries a positive electrical charge as strong as the electron's negative charge and is found in atomic nuclei.

radioactivity The spontaneous emission of particles and gamma rays by the disintegration of a nucleus. The term also refers to the emissions.

spallation A nuclear reaction that ejects light particles as the result of bombardment (as by high-energy protons).

Brush Engineered Materials
17876 St. Clair Avenue
Cleveland, OH 44110
(216) 486-4200
Web site: http://www.beminc.com

Mineralogical Museum
Irénée du Pont Mineral Room
114 Old College
University of Delaware
Newark, DE 19716-2509
(302) 831-8240; Mineralogical Museum Display Area (302) 831-4940
Web site: http://www.museums.udel.edu/mineral/index.html

Web Sites

Due to the changing nature of Internet links, the Rosen Publishing Group, Inc., has developed an online list of Web sites related to the subject of this book. This site is updated regularly. Please use this link to access the list:

http://www.rosenlinks.com/uept/bery

For Further Reading

Oxlade, Chris. *Elements and Compounds*. Chicago, IL: Heinemann, 2002.

Sacks, Oliver. *Uncle Tungsten: Memories of a Chemical Boyhood*. New York, NY: Vintage Books, 2002.

Swertka, Albert. *A Guide to Elements*. New York, NY: Oxford University Press, 2002.

VanCleave, Janice. *Chemistry for Every Kid: 101 Easy Experiments that Really Work*. New York, NY: John Wiley and Sons, 1989.

Bibliography

Brown, Andrew. *The Neutron and the Bomb: A Biography of Sir James Chadwick*. New York, NY: Oxford University Press, 1997.

Cunningham, Larry D. *Beryllium Recycling in the United States in 2000, Open-File Report 03-282*. Reston, VA: U.S. Geological Survey, 2003.

Harell, James A. "Archaeological Geology of the World's First Emerald Mine." *Geoscience Canada*, Vol. 31, No. 2, June 2004, pp. 69–76.

Kolanz, Marc E. "Introduction to Beryllium: Uses, Regulatory History, and Disease." *Applied Occupational and Environmental Hygiene*, Vol. 16, No. 5, May, 2001, pp. 559–567.

NASA. "A New Type of Mirror, a New Type of Telescope (3 of 3)." Retrieved January 10, 2006 (http://www.jwst.nasa.gov/OTE/mirrortale3.html).

Parsons, Charles Lathrop. *The Chemistry and Literature of Beryllium*. Easton, PA: The Chemical Publishing Co., 1909.

Tipler, Paul A. *Foundations of Modern Physics.* New York, NY: Worth Publishers, 1969.

Yarnell, Amanda. "Improving on Nature." *Chemical and Engineering News*, Vol. 82, No. 5, February 2, 2004, pp. 28–29.

Index

About the Author

Rick Adair is an energy policy reporter based in Seattle, Washington. After receiving his Ph.D. in earth sciences from the University of California at San Diego, he worked as a seismologist on a nuclear waste repository project and as a geophysicist on energy and ocean acoustics projects. He then provided programming and analysis for a planetary geology company whose cameras have orbited Mars. He has been a reporter since 1998, working science and environmental beats for papers in California and Nevada, before becoming a writer and news editor for an energy policy publication.

Photo Credits

Cover, pp. 1, 15, 18, 24, 25, 33, 36, 40, 42–43 by Tahara Anderson; p. 5 © Lester V. Bergman/Corbis; p. 8 © Boyer/Roger Viollet/Getty Images; pp. 10, 30 by Mark Golebiowski; p. 27 courtesy of the U.S. Geological Survey; p. 29 courtesy of NASA Marshall Space Center; p. 34 © Arnold Fisher/Photo Researchers, Inc.; p. 37 © Jonathan Blair; p. 39 © Adrian Burke/Corbis.

Designer: Tahara Anderson; Editor: Kathy Kuhtz Campbell